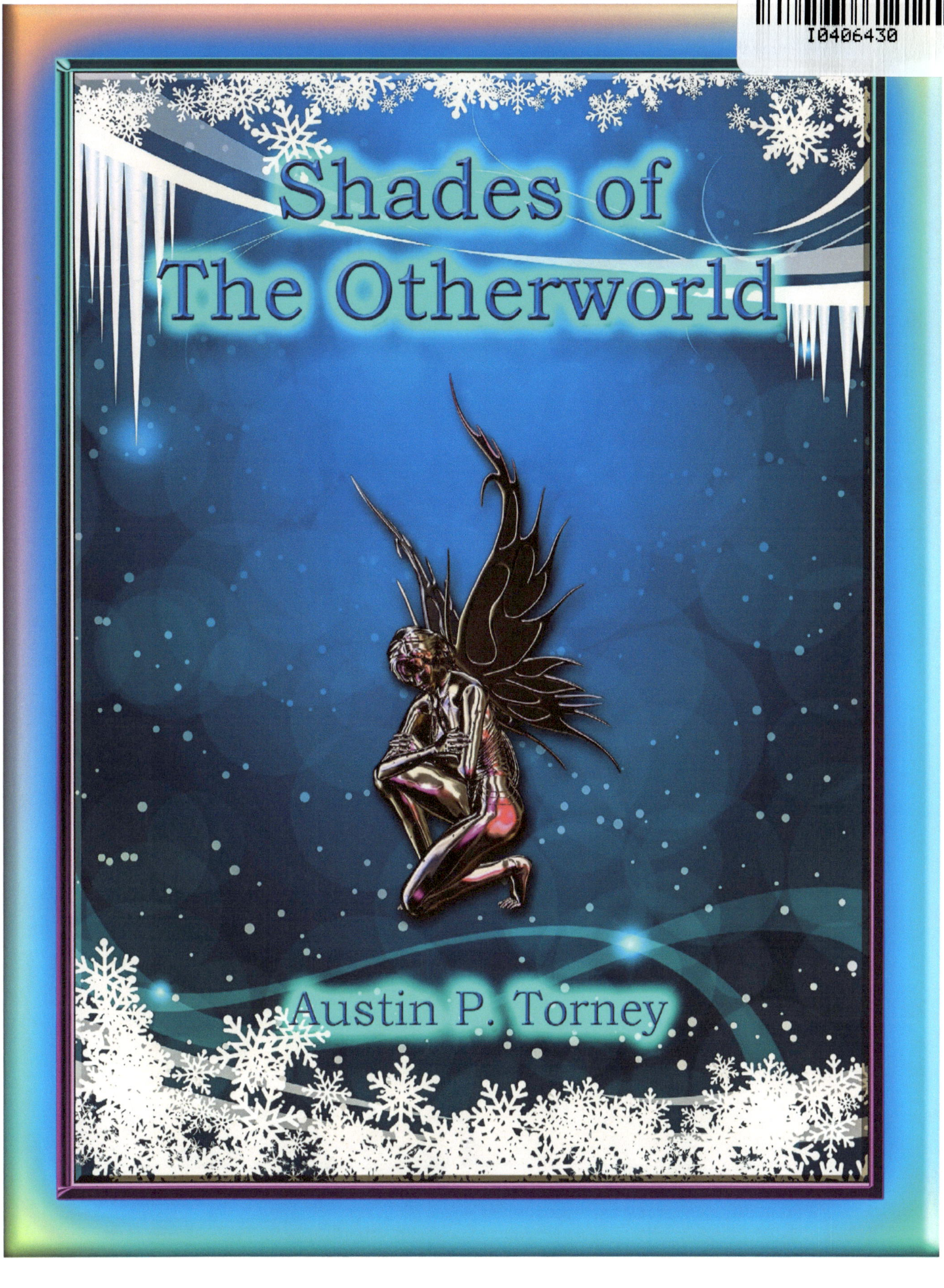

Shades of The Otherworld

Austin P. Torney

Could there be more to this world—
Those of the undrawn shades unfurled?

Is there a universe alongside this bright zone,
A parallel, twilight world overlapping our own?

Are there shadow beings all about us,
That we can only perceive as blankness?

They'd be made of but the dark matter,

Yet lively with their own kind of chatter,
These shades flowing right on through us—
We the lighted "plus" to their dark "minus".

These pale shadows of our attendants,
Are not as us—of light's extent,
But are as black clouds of a coal sack;
Nay, they're not even dark or black,
But are of an invisible bivouac.

Dark matter and its shadows traverse

The bulk of the missing-mass universe.

The shades of evening draw us on—

We must look to the past, upon the first eon.

Two distinct families of matter

Were created in the Big Freeze batter,

Just those two that did then so accrue

When they were frozen out
of the primordial stew

As the fetal universe was cooling

And the hearty gruel was ungrueling.

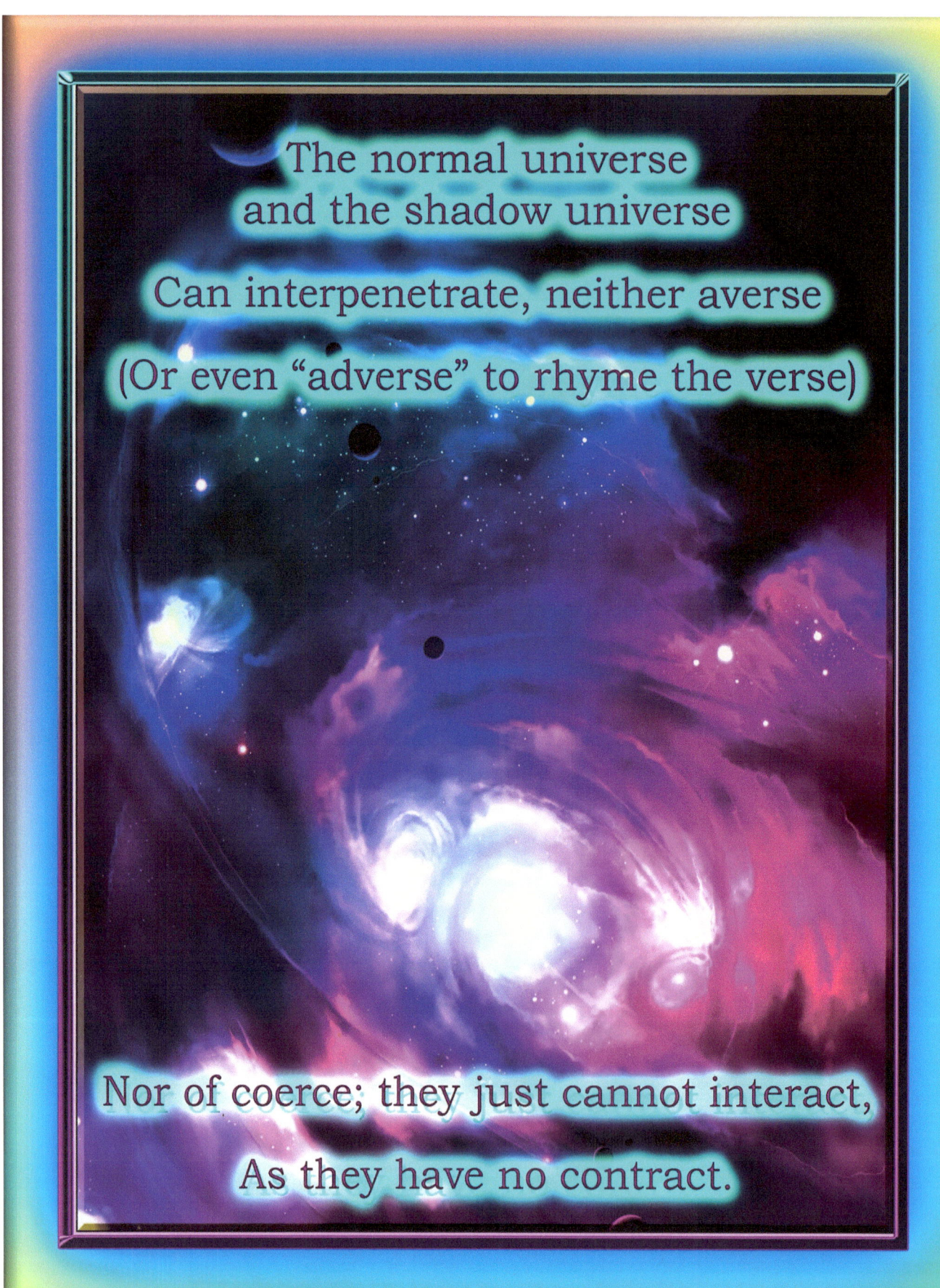

The normal universe
and the shadow universe

Can interpenetrate, neither averse

(Or even "adverse" to rhyme the verse)

Nor of coerce; they just cannot interact,

As they have no contract.

If the shadow universe was richly sown
It could have evolved along with our own.
Shadow planets could form
Around shadow stars as norms
And become populated with swarms
Of those shadow beings lukewarm.

They would be invisible specters, unseen phantoms,

Unobserved presences, indiscernible apparitions,

Imperceptible wraiths, unnoticed spirits, magic places,

Inconspicuous spooks, and hidden traces…

But first we must ask what makes a universe,

Such as ours, the one in which we immerse.

It is the forces that count for everything,

Matter being but a secondary singing,

For atoms exert forces through space,

Especially of the electromagnetic race;

So, then, it is forces that disburse

The currency of a rich universe.

This is why we don't fall through a chair—
That mostly empty space of "thin air"
When we decide to sit down there.

Space is a kind of a large-scale limitation
Of an underlying
discrete network of connections.

Atoms would not even know at all

That their companions existed, with no call,

Without the push or pull of the forces' thrall,

As then they themselves would be as pall

As some ghosts passing through a wall.

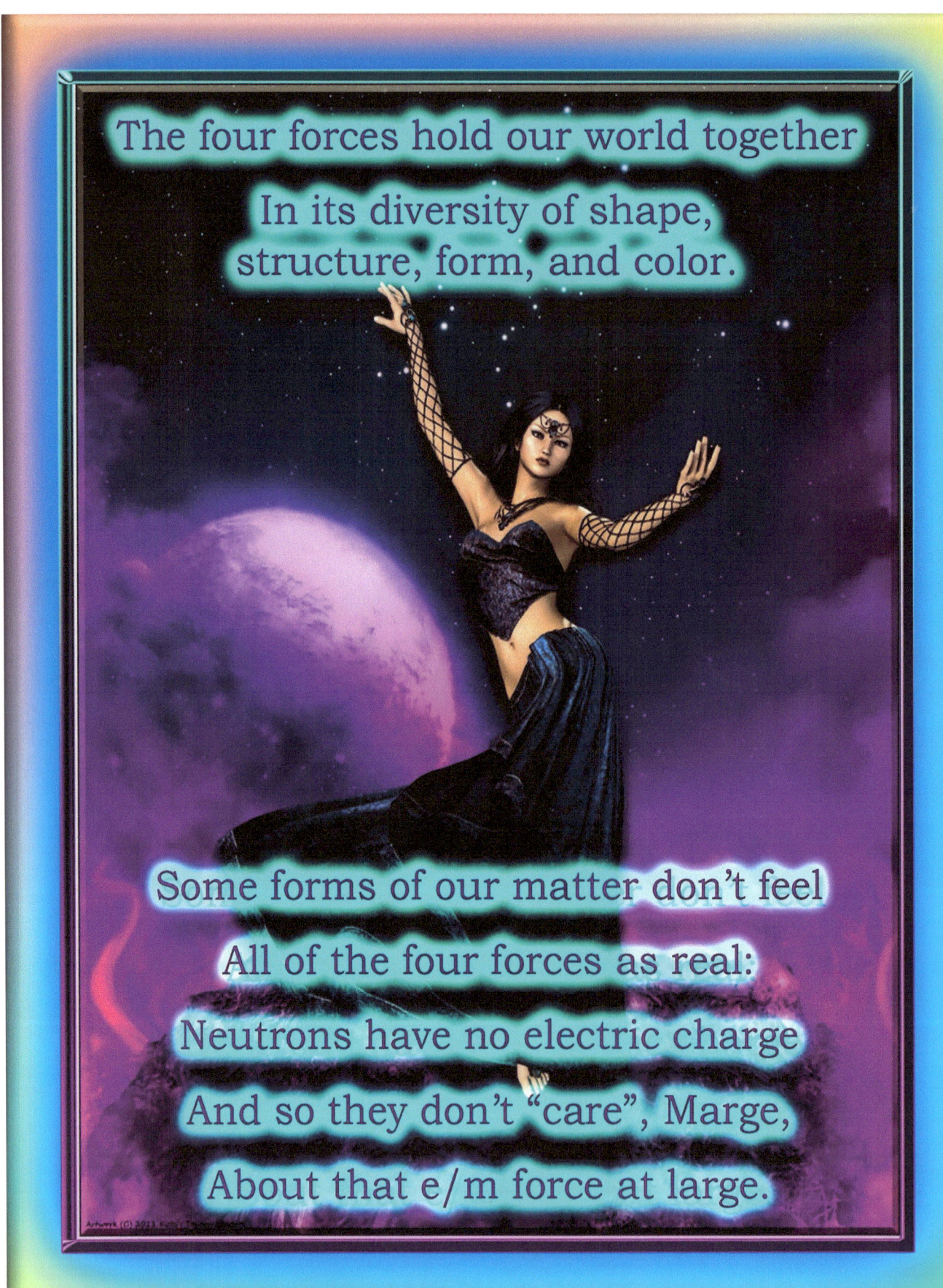

The four forces hold our world together
In its diversity of shape,
structure, form, and color.

Some forms of our matter don't feel
All of the four forces as real:
Neutrons have no electric charge
And so they don't "care", Marge,
About that e/m force at large.

Suppose some form of matter didn't feel
Any of the four forces that became real?

Dark Matter doesn't appear to discourse,
Not having the resource
of its own special forces
To bind it together; no packhorses.

All it can feel is the "force" of gravity,

And perhaps
the weak force's changeability—

Which is for decay, and not stability;

In fact, both forces are weak, a pravity.

You cannot hold a person-size lump
Of matter together with just gravity's slump;

So, then, no interesting lumps can form
In the dark universe, not even unicorns.

Even making a star or a planet

Is difficult with just
gravity alone working on it,

For the electromagnetic force is crucial

To slowing any of the material

Down enough to hold it in one place;

So, then, there can be no shadow race...

...No veiled hints, obscured suggestions,

Unknown impressions, out of sight suspicions,
Nor any supposed tinges, shimmering glimmers,
Resembling semblances, or ghostly whispers.

What has no light is but a dark shade,
With no creatures therein made.

So, dark matter is not a source for being;
'Tis but a very large matter to us unseeing.

And, yet, is it we who are the outsiders,

Our luminous bubbles of foam the riders—

The stars, planets, and us the striders—

On the vast ocean
of dark matters much wider.

We were an afterword,

Yet made possible, nonetheless,

By the dark matter—since it was oblivious

To much of the great primeval blast,

It forming filaments that could last,

Attracting our regular matter

That was everywhere splattered,

Into the pearls of the galaxies

Strung along like cosmic necklaces.

www.ingramcontent.com/pod-product-compliance
Lightning Source LLC
Chambersburg PA
CBHW040752200526
45159CB00025B/1865